Cyborg

Alphabet

Definitions of words and concepts are paraphrased and/or contextualized from both the;

Oxford University Press. (n.d.). Oxford Advanced Learner's Dictionary. Retrieved October 20, 2022, from https://www.oxfordlearnersdictionaries.com

Cambridge University Press (n.d.). Cambridge dictionary. Retrieved September 20, 2022, from https://dictionary.cambridge.org

You may contact the author for any questions or reasons by emailing feedback@sabrenetics.com

ISBN: 978-0-6456579-0-6

1st edition, Hardback,

Published January 2023

Published by Sabrenetics

www.sabrenetics.com

Acknowledgements

"For the little ones I never had the chance to meet,
who wanted to grow up and stamp your robot feet.
I dedicate this to you."

Anatomy means studying the inside and outside parts of living things like humans and animals. Anatomists are professionals that help draw pictures of the human body and different body parts like arms and legs. Professionals are people that are good at doing something. Anyone can be a professional if they keep trying to do something and learning from their mistakes. With human body pictures, we can learn more about how the body works and how to fix broken things and join things together to treat different diseases in living things.

Bionics are things made using ideas from nature. Nature is everything in our world and space not made by humans. Rocks, animals, water, trees, planets, etc. all come from nature. Cars, Computers, Toys, Cartoons etc. are all made by humans, called inventions. When you see a bird flying with its wings and you make a paper airplane shaped like the same bird, that paper airplane was invented using bionics. When you see a turtle with a hard shell and you make a car with a hard shell, that car was invented using bionics. Is there anything from nature that gives you a clever idea to make?

The word **Cyborg** is a portmanteau. Portmanteau means joining words together to make a newer word. The letters of the first syllable of Cyborg "Cyb" stand for Cybernetic which means "the scientific study of human and machine systems ". The letters of the second syllable of Cyborg "Org" stand for Organism which means "Living thing".

Cyb + Org = CYBORG

Together, the words Cybernetic Organism join to make the newer word "Cyborg."
A Cyborg is partly technology and partly living, like a half human half robot.
Cyborgs are made by adding artificial tools to living things. Someone with
hearing aids, a fake leg or pacemaker are all different Cyborgs.
Can you name any Cyborgs?

Data is any type of information we need to find out about.
Data can be physical like a piece of paper or electronic like a website.
Data can be facts, numbers, colors, sounds and anything you can think of.
When data is found and used to make something important, that data turns into information. Information is anything we find interesting and useful.
Information can be made into anything like books, videos, or songs.
So, before we start a project or homework, first we get data,
then we turn it into information.

When we want to find out information about what you are thinking of, we can do 2 things. We can ask you to talk, draw, or show us what you are thinking of. Or we can read your brain to find out what you are focusing on. Focus means the center of interest, attraction, or attention. When your eyes focus on something you see, or when your ears hear something interesting, we can read your brain waves using a method called **electroencephalography** (EEG). Electroencephalography uses something called an electrode which are tools stuck to your head that help picture what your brain is focusing on. If there is something wrong with somebody's heart, brain, muscles or body, scientists can use different methods to see what is wrong and find ways to help.

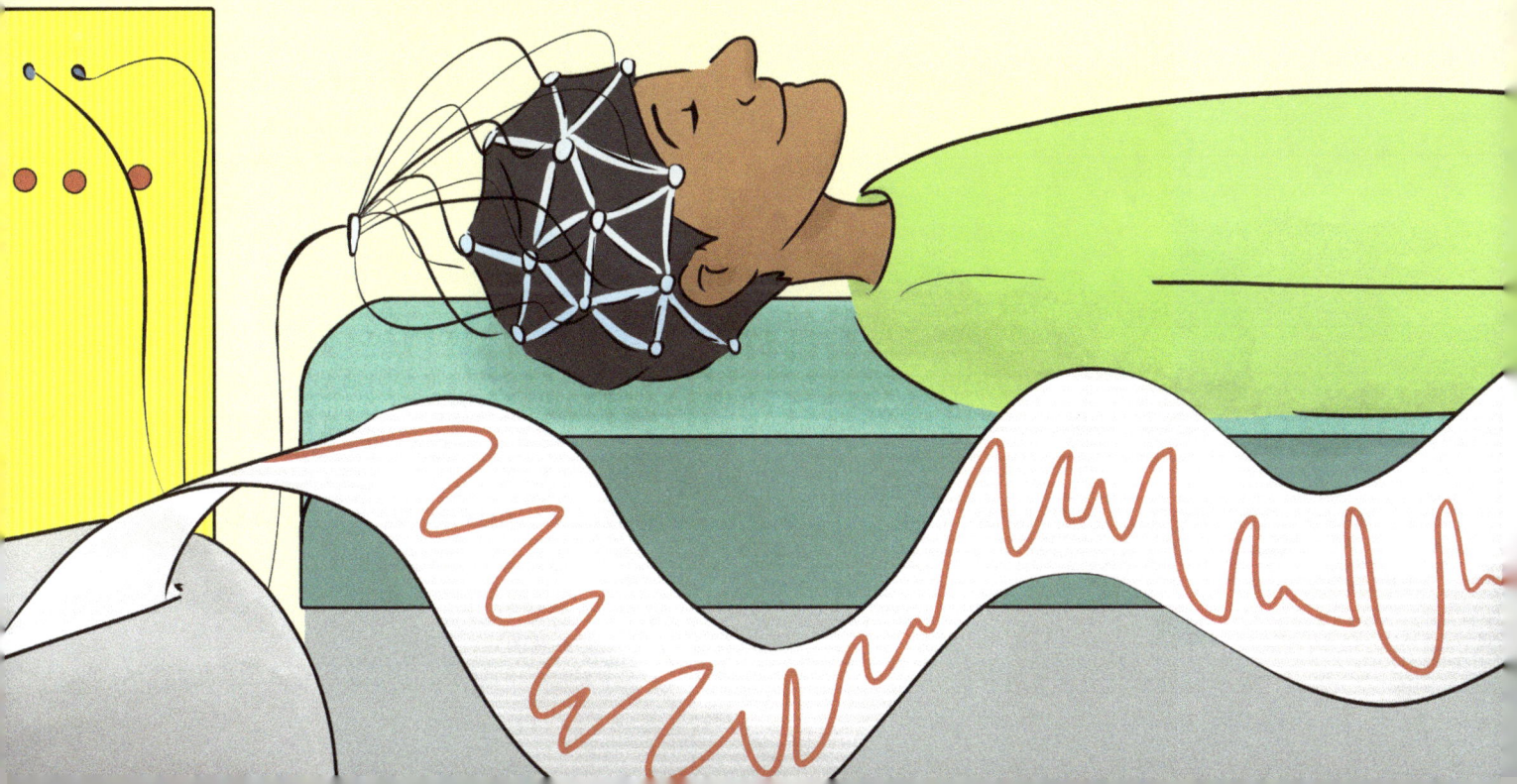

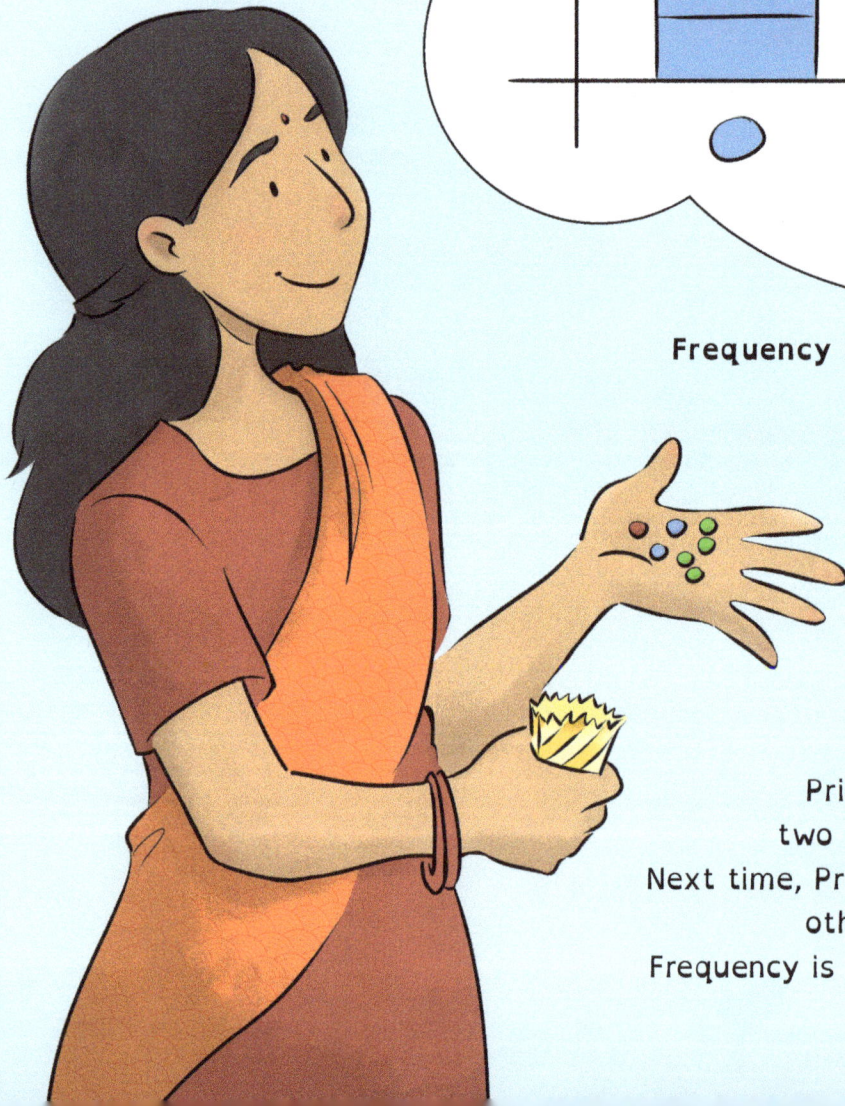

Frequency is how many times something happens with a set of data. For example, a girl named Priya tried to find red gum in a mixed Lolly bag of colors.
Priya finds
two (2) blue gums,
one (1) red gum
and four (4) green gums.
The frequency is the number of times each color came out.
Priya finds that the frequency of gums is two (2) blue, one (1) red and four (4) green.
Next time, Priya can use this information for letting others know what gums are easy to find.
Frequency is also the rate at which a wave repeats.

Gait is the way someone walks.
Walking requires balance and coordination of muscles, bones, and joints.
Balance means standing upright without falling. Coordination means working to do
something like planning to do homework. Everyone's gait is different.
Everyone walks differently because their balance and coordination can be affected by
the shape or weight of their body or differences in bones and muscles.
When people have leg injuries such as a broken leg,
then their gait is called abnormal. Not normal.

The **heart** is the central part of something. In living things like us (humans), the heart is an organ that supplies oxygenated blood to the body. Oxygenated blood carries oxygen (a gas) which is mandatory for life. An organ is a body part inside of us. We have lungs to breathe air, stomach to break down food, and brain to think. The air we breathe, called oxygen, goes into our lungs and into our blood. Blood is a red color liquid like water that leaks when we get cuts. The heart pumps the blood and oxygen all throughout our body to provide it with the energy to work.

When the heart stops, we can't pump blood and lose our life.
Life means growing, changing and multiplying. Life separates us from inorganic things
like rocks, water, walls and cars which don't have growth, can't multiply, have no hearts
and can't think. The opposite of life is death; when our bodies stop working.
When adults get lonely, we say they have a broken heart because
they lose all their energy.

Implants are anything that is put inside something else.
Implants can be made of anything such as plastic, carbon or metal.
Implants can be cosmetics which are made for looks, or reconstructive ones which are made for repairing body parts. Some examples of implants are fake bones, fake eyes, and even fake teeth. Implants help people with disabilities to improve their lives. When people get hurt and lose their body parts, they can use Implants to fix them. Biomedical Engineers are professionals that help make implants.
Implants are made to be biocompatible which means they are safe for your body.

Joints are any area where different things are joined together.
In human bodies, joints are the area where two bones meet.
In human bodies, our elbows and knees are joints where the upper and lower arm and the leg connect to. Humans need joints so we can bend our bodies. When our joints become bad, we can make artificial joints. Artificial means something made by humans instead of found in nature. When a human makes a car, that car is artificially made. When a tree grows out of the soil, that tree is made naturally.

Kilobytes are small pieces of electronic data.

Electronic data is information found in electricity. Electricity is the invisible (you cannot see) power that's food for machines like TV and gaming consoles.

Machines are artificial devices also called tools made by humans to make work easier.

Just like our power is food and water, computers use electricity for power.

When you watch TV, what you are seeing is electronic data. The TV is powered by electricity and the electricity in the TV carries electronic data called TV Shows.

Electricity is dangerous because you cannot see it, eat it and it can shock and hurt us.

Computers cannot eat food and water, so it is dangerous for them too!

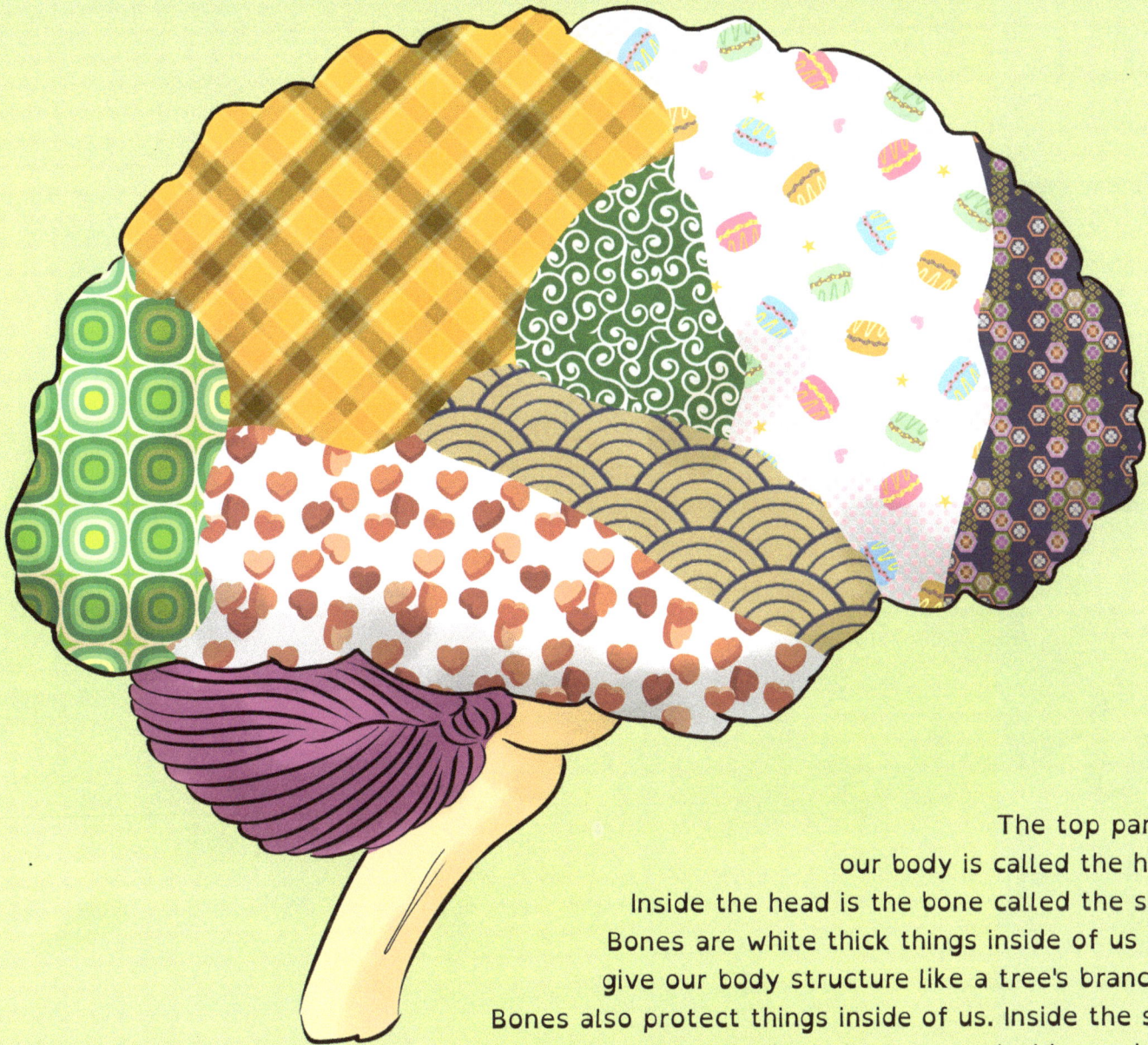

The top part of
our body is called the head.
Inside the head is the bone called the skull.
Bones are white thick things inside of us that
give our body structure like a tree's branches.
Bones also protect things inside of us. Inside the skull
is the brain. The brain is an organ inside our head.
The brain controls our body and what makes us humans.
The brain allows us to think, remember, control, and feel emotions. The biggest part of
the brain is called the Cerebrum which is made of different parts called **lobes**. Every lobe
has a different job, some allow us to remember, some to think and some to move.

Medicine is the science of diagnosis, treatment and prevention of diseases. Medicine is learning about what's wrong with living things and trying to fix them. Diagnosis means finding out a disease, condition or injury from its signs and symptoms. Signs are clues you can find out by observing something like seeing a dog's house is a sign of a dog nearby. Symptoms are physical or mental signs of a condition or disease. When you see someone who's crying, the symptom can be that they are in pain or sad. Treatment is the medical care given to a person or animal for an illness or injury. Treatment helps in healing from a disease. When someone is bleeding and hurt from running and falling over, the treatment is to use a bandage and sleep. Diseases are harmful variations from the normal state of an organism. When your body is normal, you are in good health. When your body feels bad or wrong, or something is different, there could be a disease involved.

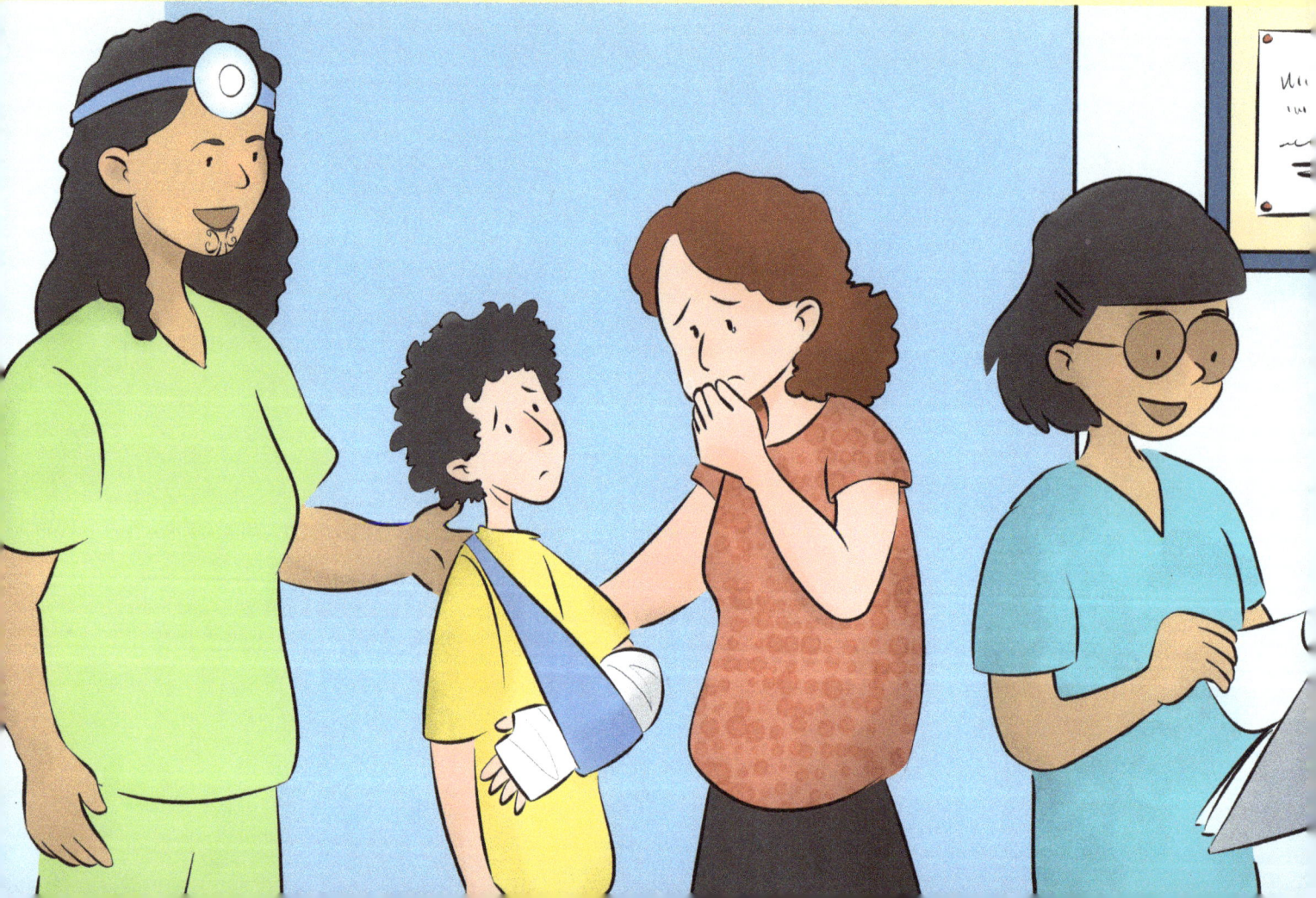

Disease usually occurs due to germs or changes in the body's function. Germs are small living things that can harm humans and animals. To see germs with our eyes, we use a microscope. Prevention means stopping something from happening. When someone eats too much, prevention is telling them to stop eating too much. Washing your hands with soap can prevent you from diseases and germs. Doctors are scientists whose job is to help fix any issues your body has by finding out what's wrong, coming up with a plan to fix you, then giving you medication, advice and possibly surgery to fix you. Medication is a treatment using drugs. Drugs can be food, water, and things natural or artificial that are good to eat and make you feel better. Surgery is the treatment of diseases, conditions or injuries by taking out, fixing, or replacing things inside organisms. When you have a loose tooth, a special doctor called a dentist will take it out in surgery.

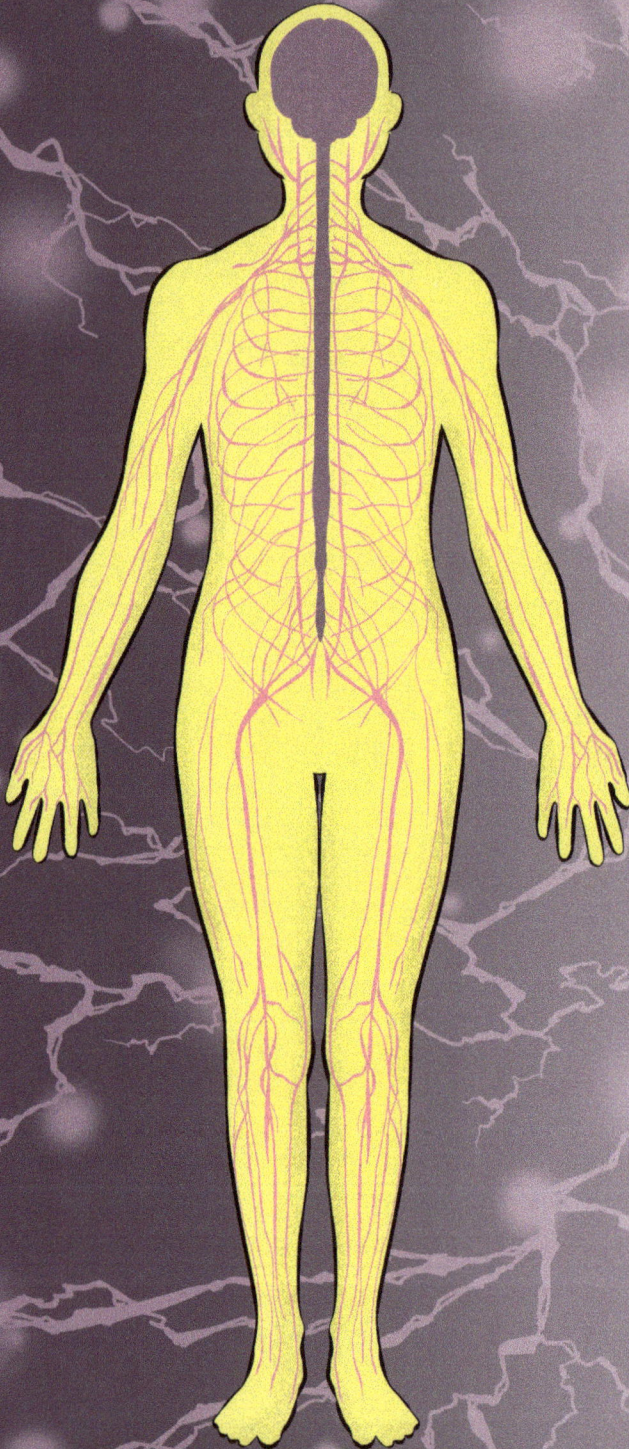

Living things are made of different things working together.
Cells are the smallest structure of a living thing. Living things like humans and animals are made up of billions and trillions of cells.
If a house is made of wood, furniture, lights and Toilets,
The human body is made of cells, tissues, and organs all working together in systems.
When cells combine, they form tissues, and when tissue combines, they form organs. There are brain cells in your head that help you think.
There are red blood cells that move like water inside of you to transport energy (power) and waste (rubbish).
Nerves are cells that receive and send signals between the body and the brain. If you touch something hot and burn your hand, the nerve endings send a message to your brain saying what you touched is hot, so you know to stop touching it.
When you walk, run, jump or move, your nerve cells receive messages from the brain telling it what to do.
When certain nerves are damaged, we lose the ability to function.
Sometimes people use wheelchairs instead of walking because their nerves are damaged.

Organisms are individual living things. When different organs combine, they form an organ system, and when organ systems work together, they form an organism. Organisms can be humans which are us, plants like trees or animals like dogs. Individual means self. You are an individual, your parents, teachers, friends, and so on are individuals as well. Every individual is different. Some individuals like blue, some like pink. Some individuals look brown, others white and black. The biological taxonomy is a chart of all living things and how they were all related. Like a big family tree. Charts are pictures filled with information about something. Like the alphabet or numbers. All living things are related by characteristics. Related means being part of a family. For example, mum, dad, brother and sister are all related. Characteristics mean things that describe something such as how many bones we have, where we live in the sea or mountains, or what food we eat. Even though we may look or sound different, we are all related together like a big family.

Prostheses are artificial body parts.
Sometimes people have accidents that damage their body or are born with missing or malformed body parts. Malformed means something isn't what you expect; like only having half a leg or a hand with 6 fingers. Prostheses are made for helping damaged, missing, or malformed parts of the human body. Fake arms, legs, fingers or fake teeth are all examples of Prostheses. Professionals called Biomedical Engineers help make Prostheses. They first find the anatomy of your body using tools like x-rays to draw your missing, damaged or abnormal body parts like missing legs. Then they make a prosthesis to either replace or aid that abnormal body part like creating a fake leg.

Quacks are fraudulent people that claim to be experts in something. Fraudulent means saying you are good at something when you are not. Experts are professionals at something like painting or teaching. Doctors are scientists who are professionals and experts at helping sick people feel better. Some people pretend to be Doctors without reading and learning. Sometimes people lie so people will like them, or they want something. Lying means not telling the truth. Lying is bad, dangerous and has consequences. If we lie about something, people can get hurt which is called bad consequences. We should always learn the truth from real scientists, teachers and doctors. You can tell someone's an expert when they experiment, learn, don't lie and can teach, make or say things that other honest experts do.

Rehabilitation is the process of restoring someone's health, reputation or privileges through training and therapy. Health is how good we feel, and our bodies are. If we are hurt in an accident, our health suffers. If we get better, then our health gets better. Reputation is what people think of us. Do others like you, do they hate you? If you're a bad person, your reputation will be bad because people don't like you. If you are a good person, you will have a good reputation. Privilege is what we are allowed or able to do. If we can run around outside safely, we are privileged to live in a safe country.

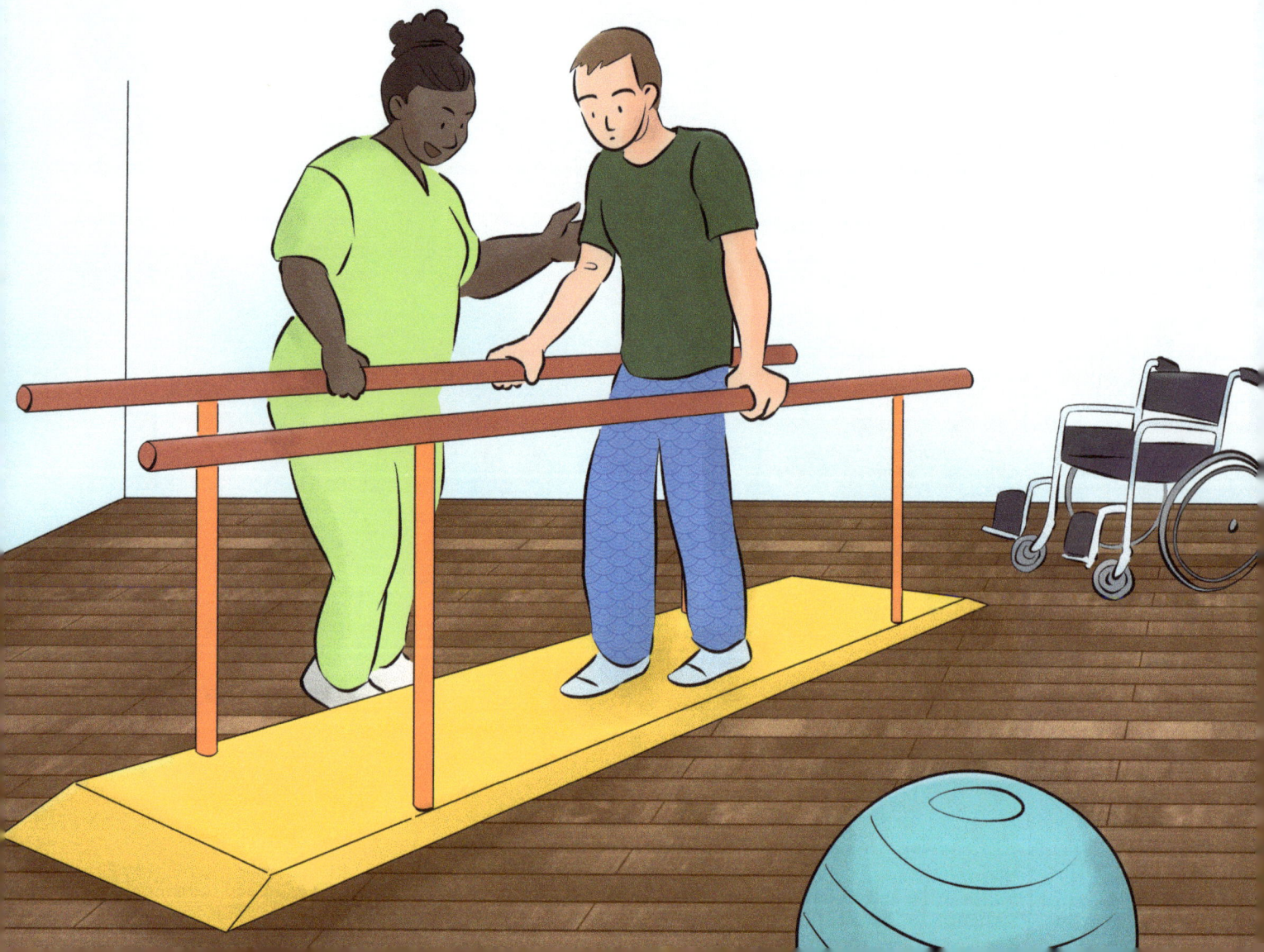

In poor dangerous countries, people don't have the privilege to be safe outside because bad people will hurt them. Therapy is the treatment (way) for making a disorder (problem) better. If you eat too much food, therapy can be eating smaller food all throughout the day, so you feel fuller and less hungry. Training is teaching a person or animal a skill or a specific behavior. When you teach a dog to sit, shake hands and fetch a ball, you are training the dog. Rehabilitation is a process used to fix people for many reasons like helping people stay out of jail, stop taking drugs, or correct their health so they can move and function like normal.

Science is knowledge about the world based on facts, evidence, and experiments. There are different branches of science such as biology, physics, and chemistry. Knowledge is what you know. If you like playing a fun game, you have knowledge of which game is fun to play in the future. To know something is to have knowledge about it. Facts are true things. If it's feeling hot and it's a hot day, the fact is that today is a hot day. Fiction is fake stories like comic books, cartoons, and lies. Evidence is a sign that something is a fact. When you find a bird flying in the sky, you find evidence that birds fly in the sky.

Experiments are to find out if something is true or false. When you see lots of different birds flying in the sky, you do an experiment to prove birds fly by waiting and watching the birds. Scientists are humans whose job is to do experiments to find evidence to see what's fact or fiction. Jobs are things you keep doing to learn something, make something or get something in return. Humans are people like us, yourself, parents, teachers and friends are all humans. Animals are non-human living things. Good examples include pets, dogs, cats, fish, and insects. Scientists help us learn about the world, so we can read all about it and know what's fact and fiction.

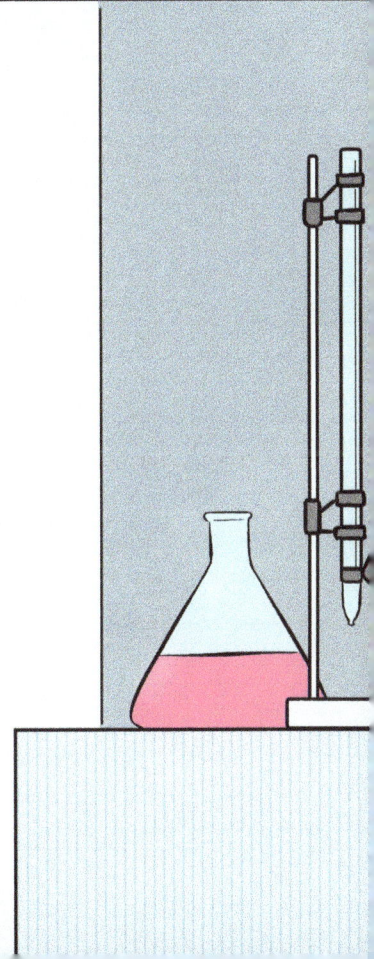

Transhuman are halfway between being a normal human person and post-human. Transhuman can be people like cyborgs which are partly human and partly robot. Transhuman can also be superheroes from cartoons, comics, games, shows, and movies. Posthumans are humans that have changed so much that they're not recognizable as people anymore. Recognizable means knowing what something is or looks like from knowledge. A museum is a place where they keep old stuff to see. If you read a dinosaur book and see something that looks like a dinosaur at the museum, you can recognize the dinosaur from the book. Post-humans aren't real, we can only imagine them. In the future, a very long time from now, humans will change so much that we won't Know what we're going to look like.

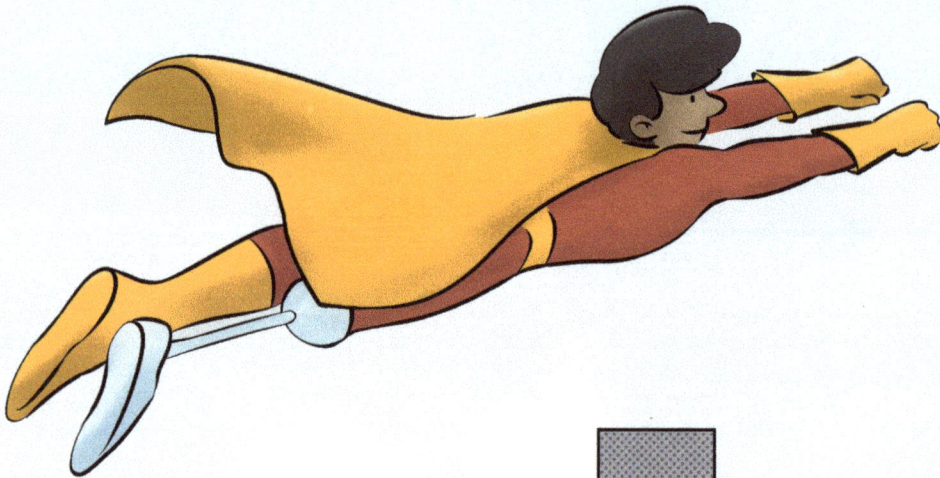

Utilitarianism is a moral thought.

Morals are what we think is right or wrong. Should we steal or help? Should we lie or tell the truth? These are examples of moral questions. Utilitarianism means that actions are right if they are found useful or for the benefit of a majority. If a car accident happens and two (2) cars hit each other and the first car has only one (1) old man while the second car has a family of four, dad, mum, and two (2) children both brother and sister, should we help the old man first? Or the family first?

Utilitarianism says we should help the family because there are more people to help.

Every day, we use our morals to find what is the good thing or bad thing to do.

Some people follow Utilitarianism, and some people have their own morals they follow.

Electricity is the invisible power that keeps machines on. Machines are Computers, TVs, and Fridges which are all tools made to make life easier or fun. Inside machines are circuits. Circuits are made of semiconductors and tools. Electricity travels through the circuit like a river going in a circle. When electricity travels one way down the circuit, it's called a current, like a river.

When electric current runs down the circuit, it powers different semiconductors and tools like light bulbs and TV, and computer screens. Electric current is like blood in a human. The heart pumps the blood all over the body so we can get nutrient energy to live.

In a machine, the electric current travels down the circuit to give the machine electric energy to live.

Voltage is how strong the current is. In the human body, the heart pushes the blood all over the body; that push is like a voltage for machines. It is how strong the push of current is.

Signals are actions and sounds used to pass information.
Examples of Signals are smoke signals, hand signs or traffic
lights. Signals move through **waves**. Waves are the way that
signals travel. Waves can be sound, radio, or even light.
A wave is the energy that travels back and forth
in a medium like the sea moving back and forth creating waves.
Talking to someone back and forth creates sound waves.
A signal could be talking to someone,
and the wave used to carry that signal is called sound.

Scientists use **x-rays** to take a photo called an x-ray image of the inside of the human body. The X-ray machine uses invisible waves called x-rays that can go straight through the body. When the X-rays go inside the body, they create a picture of the inside of you. Bones, the inside soft parts and air can be seen on the X-ray images in white, shades of grey, and black colors. X-rays help doctors to diagnose a disease. When someone breaks a bone in the body like in an accident a doctor will take an X-ray image of that part to see what happened inside of them. If someone accidentally swallows something they shouldn't like a toy or bone, x-rays can help see inside of you. Doctors use x-ray images to learn what's wrong and plan to fix it.

Youth is the age of life when you are still young. Being young, you are still growing. Once you are grown up, you stop being young and lose your youth. When you are young, you have more energy, which is the power to move and are fast, and strong. When you are older, you become less energetic and become more tired, slower, and weaker. Scientists discover new ways to stay younger and live longer by telling us what is healthy, like eating vegetables and playing under the sun. If we do not listen to our doctors, we damage our health, which is why it is best to listen to our doctors to keep being as youthful as possible.

In science, a man is called a "male" and a woman is called a "female". "male" and "female" are both called sexes. Lots of animals have different sexes; there are "male" dogs and "female" cats. Some animals like plants can be of both sexes. A **zygote** is a new life created when a male and female's body fluids (sperm from male and egg from female) join. The zygote lives in the female's body until it gets older and then turns into a baby.

www.ingramcontent.com/pod-product-compliance
Lightning Source LLC
Chambersburg PA
CBHW040250100426
42811CB00011B/1210